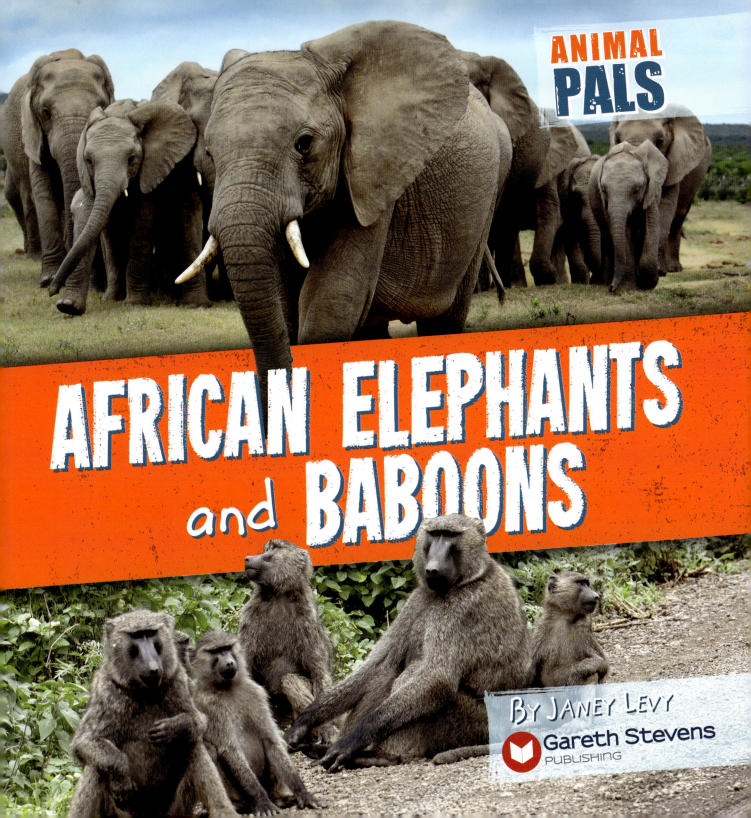

Please visit our website, www.garethstevens.com. For a free color catalog of all our high-quality books, call toll free 1-800-542-2595 or fax 1-877-542-2596.

Library of Congress Cataloging-in-Publication Data

Names: Levy, Janey, author.
Title: African elephants and baboons / Janey Levy.
Description: New York : Gareth Stevens Publishing, [2022] | Series: Animal pals | Includes index.
Identifiers: LCCN 2020031049 (print) | LCCN 2020031050 (ebook) | ISBN 9781538266663 (library binding) | ISBN 9781538266649 (paperback) | ISBN 9781538266656 (set) | ISBN 9781538266670 (ebook)
Subjects: LCSH: African elephant–Juvenile literature. | Baboons–Juvenile literature.
Classification: LCC QL737.P98 L485 2022 (print) | LCC QL737.P98 (ebook) | DDC 599.67/4–dc23
LC record available at https://lccn.loc.gov/2020031049
LC ebook record available at https://lccn.loc.gov/2020031050

First Edition

Published in 2022 by
Gareth Stevens Publishing
29 E. 21st Street
New York, NY 10010

Copyright © 2022 Gareth Stevens Publishing

Designer: Andrea Davison-Bartolotta
Editor: Monika Davies

Photo credits: Cover (top) Michael Potter11/Shutterstock.com; cover (bottom) Vladislav T. Jirousek/Shutterstock.com; pp. 5 (top), 7, 11 Wolfgang Kaehler/LightRocket via Getty Images; p. 5 (bottom) Auscape/Universal Images Group via Getty Images; p. 8 mariait/Shutterstock.com; p. 9 Bernd Wasiolka/McPhoto/ullstein bild via Getty Images; p. 13 Tim Graham/Getty Images; p. 14 Melissa Valente/Shutterstock.com; pp. 15, 16 pingebat/Shutterstock.com; p. 17 Eric Lafforgue/Art in All of Us/Corbis via Getty Images; p. 19 Roger de la Harpe/Education Images/Universal Images Group via Getty Images; p. 20 Holly Auchincloss/Shutterstock.com; p. 21 Pascale Gueret/Shutterstock.com.

All rights reserved. No part of this book may be reproduced in any form without permission in writing from the publisher, except by a reviewer.

Printed in the United States of America

CPSIA compliance information: Batch #CSGS22: For further information contact Gareth Stevens, New York, New York at 1-800-542-2595.

CONTENTS

African Friends 4

Awesome African Elephants 6

Follow the Leader 8

Outstanding Olive Baboons 10

Baboon Life Basics 12

Where Elephants and Baboons Bond 14

About Eritrea 16

The Special Elephant-Baboon Bond 18

More Mutualistic Relationships 20

Glossary 22

For More Information 23

Index 24

Words in the glossary appear in **bold** type the first time they are used in the text.

AFRICAN FRIENDS

Good times are better when you share them with someone else. Hard times are easier when you team up with one another. That's also true in the animal world! Take African elephants and olive baboons. You might never imagine them together. But they help each other get through hard times.

These animals make a strange pair. African elephants are the largest living land animals. Olive baboons are monkeys that might weigh about the same as you do! Inside this book, you'll learn more about the animals and their unusual bond.

FACT FINDER

African elephants and olive baboons have a mutualistic **relationship**. This is a relationship between two different kinds of animals that benefits both of them.

Elephants and olive baboons don't hang out with each other all the time. But they get together when they can be most helpful to each other.

AWESOME AFRICAN ELEPHANTS

African elephants are huge! They can be up to 13 feet (4 m) tall and weigh up to 14,000 pounds (6,350 kg). Their trunk alone can weigh up to 310 pounds (140 kg)! Elephants use their trunk to breathe, drink water, and spray water over their body to cool off. The trunk also has two "fingers" on the end that are used to pick up objects.

Both male and female elephants have **tusks**. They use them for digging, lifting, fighting, and collecting food.

FACT FINDER!

Sadly, African elephants' ivory tusks can put them in danger. Some humans illegally hunt elephants for their tusks because of the huge value placed on ivory.

The tusks of a male African elephant can weigh over 220 pounds (100 kg)!

7

FOLLOW THE LEADER

African elephants live in herds made up of female family members and their young. The oldest female leads the herd. She knows where to find food and water.

African elephants are herbivores, or plant eaters. They eat leaves, fruits, grasses, roots, branches, and bark. Since these elephants are so big, they eat a lot. An adult can eat up to 300 pounds (136 kg) in one day! African elephants also need to drink lots of water. They can drink up to 50 gallons (189 L) each day.

FACT FINDER

Adult male African elephants live alone or with a few other adult males.

Herds may have anywhere from about 10 to 70 members.

OUTSTANDING OLIVE BABOONS

There are five species, or kinds, of baboons, but the olive baboon species is found in the most countries. They got their name because their fur looks gray-green—or olive green—from far away. However, each hair is made up of many colors, including yellow, brown, and black.

Male olive baboons are bigger than females. They're about 28 inches (71 cm) tall at the shoulder when standing on all fours. They usually weigh about 55 pounds (25 kg), but they can weigh up to 110 pounds (50 kg).

Olive baboons are also known for their long, doglike **snout**.

FACT FINDER!

While olive baboons have tails that can measure up to 22 inches (56 cm) long, they aren't able to use their tail to grab branches or hold onto things.

BABOON LIFE BASICS

Olive baboons spend most of their time on the ground. They live in groups, called troops, made up of a few males plus many females and their young. The females form the heart of the troop.

Olive baboons are **omnivores**. They eat grass, leaves, berries, roots, seeds, flowers, and bark. The baboons also eat bugs, spiders, and small animals such as hares and **rodents**. They're also smart enough to work as a team to hunt **prey** such as small **antelopes**.

FACT FINDER!

Predators such as lions and leopards hunt olive baboons. The baboons climb up trees to watch out for these animals.

An olive baboon troop can have as many as 150 members.

WHERE ELEPHANTS AND BABOONS BOND

African elephants live scattered across many countries in Africa, all south of the Sahara Desert. Some elephants live in **savannas**, while others live in forests. Olive baboons are found from one side of Africa to the other, mainly in countries around the **equator**.

African elephants and olive baboons live in many of the same places. But it's in the eastern African country of Eritrea that they've been seen to have a mutualistic relationship.

This map shows the places where African elephants and olive baboons live.

15

ABOUT ERITREA

Eritrea is a small country on the eastern coast of Africa, located beside the Red Sea. The country faces many challenges, or problems, that make life hard for people and animals alike.

Eritrea doesn't get much rain and suffers from **drought**. The lack of rain means much of the land is becoming dry and desertlike. The dry land makes it hard for people in Eritrea to grow food. This also means animals struggle to find food and water.

Eritrea used to be part of the country of Ethiopia. It became an independent country in 1993.

THE SPECIAL ELEPHANT-BABOON BOND

African elephants drink a lot of water. Sometimes, elephants must find water below the land's surface. They dig huge holes in the ground using their powerful feet, tusks, and trunk.

In Eritrea, observers have seen African elephants share their water with olive baboons. The baboons also help out the elephants. Sitting high in the trees, olive baboons warn the elephants when predators are near. This warning system helps the elephants stay safe. These two different animals work together as friends.

African elephants and olive baboons both benefit from their mutualistic relationship.

FACT FINDER!

Olive baboons need to drink water just like the elephants. But they aren't strong enough to dig the kind of holes the elephants can dig to get water.

MORE MUTUALISTIC RELATIONSHIPS

African elephants and olive baboons have established a special bond in Eritrea. But African elephants have other mutualistic relationships as well. It's believed that birds called cattle egrets also team up with elephants. The birds sit on top of African elephants, eating the elephants' **parasites**. The cattle egrets get a good meal and a ride to new places, while elephants get rid of their parasites.

Mutualistic relationships help improve the lives of both animals. It's the best kind of friendship in the animal kingdom!

African elephants are huge animals—and they also play a huge role in helping other living things survive in their **environment**.

GLOSSARY

antelope: an animal much like a deer that lives in Africa and southwest Asia

drought: a long period of very dry weather

environment: the natural world in which a plant or animal lives

equator: an imaginary line around Earth that is the same distance from the North and South Poles

omnivore: an animal that eats both meat and plants

parasite: a living thing that lives in, on, or with another living thing and often harms it

prey: an animal that is hunted by other animals for food

relationship: a connection between two living things

rodent: a small, furry animal with large front teeth, such as a mouse or rat

savanna: a grassland with scattered patches of trees

snout: an animal's nose and mouth

tusk: a large tooth that curves up and out of an animal's mouth

FOR MORE INFORMATION

Books

Bell, Samantha S. *The Amazing Social Lives of African Elephants.* Mankato, MN: Childs World, 2018.

McAneney, Caitie. *African Elephant: The Largest Land Mammal.* New York, NY: PowerKids Press, 2020.

Murray, Julie. *Baboon Troop.* Minneapolis, MN: ABDO Kids, 2019.

Websites

African Elephant
kids.nationalgeographic.com/animals/mammals/african-elephant/
Discover more interesting facts about African elephants on this website.

African Elephant
kids.sandiegozoo.org/animals/african-elephant
Hear an elephant's roar and learn more about the world's largest land mammal here.

Baboon
kids.britannica.com/kids/article/baboon/352811
Find out more general information about baboons here.

Publisher's note to educators and parents: Our editors have carefully reviewed these websites to ensure that they are suitable for students. Many websites change frequently, however, and we cannot guarantee that a site's future contents will continue to meet our high standards of quality and educational value. Be advised that students should be closely supervised whenever they access the internet.

INDEX

Africa, 14, 15, 16
bark, 8, 12
berries, 12
branches, 8, 11
cattle egrets, 20
drought, 16
equator, 14
Eritrea, 14, 16, 17, 18, 20
Ethiopia, 17
flowers, 12
food, 6, 8, 16
forests, 14
fruit, 8
fur, 10
grasses, 8, 12
herbivores, 8
herds, 8, 9

leaves, 8, 12
monkeys, 4
mutualistic relationship, 4, 14, 19, 20
omnivores, 12
predators, 12, 18
Red Sea, 16
roots, 8, 12
Sahara Desert, 14
savannas, 14
seeds, 12
snout, 11
species, 10
tail, 11
troops, 12, 13
trunk, 6, 18
tusks, 6, 7, 18
water, 6, 8, 16, 18, 19